BLACK HOLES AND THE FRONTIERS OF SPACE AND TIME
Inside the Cosmic Abyss

Exploring Singularities, Wormholes, and the Wonders of the Universe's Most Mysterious Phenomena

Scott W. Diego

Table of Contents

Introduction

Black holes have captivated the human imagination for centuries. They represent a profound mystery—a kind of cosmic riddle that defies easy answers and stretches the boundaries of science as we know it. For scientists, black holes pose some of the most challenging questions in the study of the universe, acting as gateways to unexplored realms of physics. They challenge our understanding of space, time, and the very fabric of reality, inviting us to look beyond the boundaries of what we can see and measure.

The allure of black holes isn't limited to scientists; they have a way of capturing the minds of anyone who's ever looked up at the night sky and wondered what might lie beyond. Unlike other celestial objects—planets, stars, or even comets—black holes seem to defy logic. Here are objects so dense that their gravitational pull can capture even light, drawing it into an unknown and seemingly unreachable depth. It's this idea of an inescapable void, a place where conventional rules break down, that feeds our curiosity. Are they merely regions of intense gravity, or could they be something more, perhaps a bridge to other parts of the cosmos or even different universes?

At the core, black holes are born from the explosive deaths of massive stars. When a star has expended its fuel and collapses under its own gravity, it leaves behind an incredibly dense core. This collapse can result in a black hole, a place where gravity becomes nearly infinite. But while we know a fair amount about how black holes form, the details of what happens within them remain shrouded in mystery. Scientists use theories and models to peer into this "cosmic abyss," but as of now, no one truly knows what lies inside or what fate befalls anything that crosses the boundary known as the event horizon.

The pursuit of knowledge about black holes is fueled by a mixture of scientific curiosity and existential wonder. Every discovery adds a new layer to our understanding, while raising even more questions. Could black holes hold the answers to the mysteries of the universe, or are they simply reminders of the limits of human knowledge? This journey into the unknown drives researchers, sparking debates and speculations about everything from the nature of reality to the possibilities of time travel. Black holes are not just astronomical phenomena; they are cosmic puzzles that inspire us to push beyond the limits of science and explore the edges of what we understand about existence itself.

Chapter 1: The Life and Death of Stars

The story of a black hole begins with the birth and evolution of a star. Stars are massive spheres of burning gases, primarily hydrogen, that shine by generating energy through nuclear fusion. Deep within a star's core, extreme temperatures and pressures ignite this fusion process, forcing hydrogen atoms to combine and form helium. This process releases immense amounts of energy in the form of light and heat, which radiates outward, balancing the force of gravity trying to pull everything inward.

As a star ages, it continues to fuse heavier elements, moving from hydrogen to helium and eventually to carbon, oxygen, and so forth. This fusion process creates a balance—outward pressure from the core pushes against the force of gravity, holding the star stable. This delicate equilibrium can sustain a star for millions or even billions of years, depending on its size. Larger stars burn through their fuel faster and live shorter, more intense lives, while smaller stars burn more slowly, glowing steadily over vast stretches of time.

When a massive star exhausts its nuclear fuel, it can no longer sustain the outward pressure needed to counteract gravity's pull. This loss of balance causes the star to collapse in on itself. The outer layers are often blown away in a powerful explosion called a supernova, a brief but dazzling event that sends shockwaves and stellar remnants across the cosmos. What remains is the core, left exposed and vulnerable to the overwhelming force of its own gravity.

For most stars, this core collapse results in a neutron star—a highly dense object composed almost entirely of neutrons, held in check by neutron degeneracy pressure. However, if the original star was exceptionally massive, even this pressure cannot counteract the force of gravity, and the core continues to collapse until it forms a black hole. In this state, gravity pulls inward with such intensity that it creates a region where not even light can escape, and space-time itself is stretched beyond recognition.

In essence, black holes represent the end stage of the life cycle for the universe's most massive stars, a final transformation that leaves behind an invisible remnant of unimaginable density and gravitational power. They are born from both destruction and creation, products of stellar death that continue to influence and reshape the cosmos around them. The formation of black holes is as much a story of cosmic life cycles as it is of the intense, unseen forces at work in the universe. Each black hole stands as a testament to the powerful processes that govern star formation, the balance of forces within, and the ultimate collapse of some of the universe's most awe-inspiring celestial bodies.

Gravity and fusion are the twin forces that shape a star's life. At its core, a star exists in a state of constant tension between the inward pull of gravity and the outward push generated by fusion reactions. Gravity tries to compress the star, pulling all of its matter inward toward the core. Meanwhile, the immense heat and pressure at the core drive nuclear fusion, a process that forces lighter atomic nuclei, primarily hydrogen, to fuse into helium, releasing energy. This energy radiates outward, creating a powerful force that counteracts gravity's inward pull, maintaining the star's stability.

As long as hydrogen fuel is plentiful, this balance persists. For millions or even billions of years, gravity and fusion hold each other in check, allowing the star to shine steadily in the cosmos. But over time, hydrogen in the core becomes depleted, and the star begins fusing heavier elements in sequence—helium to carbon, then oxygen, silicon, and so on. Each stage of fusion requires greater heat and pressure, with the core steadily growing hotter and denser.

This progression reaches a critical point when the star begins to produce iron. Fusion reactions involving iron do not release energy; instead, they absorb it. The energy balance that has sustained the star for so long is now disrupted. Without a steady outward push, gravity begins to win. The core can no longer resist the force pulling it inward, and it rapidly collapses under its own weight. This collapse sends shockwaves through the star's outer layers, often resulting in a supernova explosion that ejects much of the star's material into space.

In the most massive stars, this collapse goes a step further. With no internal pressure capable of resisting gravity, the core continues to shrink until it forms a singularity—a black hole. In this state, gravity is so intense that nothing, not even light, can escape. The outward force of fusion, which once held gravity in check, has vanished entirely, leaving behind a region of pure gravitational pull.

The intricate dance of gravity and fusion is thus both the life and death of a star. While fusion keeps gravity at bay for most of a star's life, the inevitable breakdown of this process leads to the dramatic collapse that can ultimately create a black hole. This cosmic cycle of balance and breakdown underscores the profound and powerful forces that shape the life cycle of stars, transforming them from bright beacons in the sky into some of the most enigmatic objects in the universe.

Chapter 2: What Exactly Is a Black Hole?

A black hole is defined by its intense gravitational pull, a force so overwhelming that once anything crosses a certain boundary, it can never return. This boundary, known as the event horizon, marks the point of no escape; beyond it, the gravitational force is so extreme that not even light, traveling at nearly 300,000 kilometers per second, can break free. This unique characteristic renders black holes invisible against the dark fabric of space—they appear as voids rather than as traditional celestial objects. Their presence is often inferred by the way they interact with surrounding matter or bend light around them.

The nature of a black hole stems from three primary factors: mass, density, and gravitational pull. Mass refers to the amount of matter within the black hole, which is often incomprehensibly vast, even though black holes themselves can be surprisingly compact. Density, on the other hand, describes how tightly this mass is packed within a tiny space. A black hole's core, known as a singularity, is a point of infinite density where all its mass is compressed into an infinitely small space. This concentration of mass leads to an immense gravitational pull that alters the surrounding fabric of space and time, creating the defining "trap" for light and matter.

Einstein's theory of general relativity provides a framework for understanding how black holes form and why they have such powerful gravitational fields. According to his theory, any object could theoretically become a black hole if compressed to a small enough volume relative to its mass. For example, if Earth were reduced to the size of a small marble, its density would increase so drastically that it would collapse into a black hole, with gravity preventing anything within reach from escaping. This theoretical principle underscores the unique relationship between mass, density, and gravity: it is not the amount of matter alone that creates a black hole, but the density of that matter packed into a minimal space.

In essence, black holes represent a limit of physical laws as we understand them. They are objects where gravity dominates so fully that it warps not only the behavior of nearby objects but also the very structure of space-time. Within this extreme environment, the usual rules governing mass, energy, and even time begin to break down, creating a region where traditional physics no longer applies. This makes black holes not only fascinating but also profoundly mysterious, as they push the boundaries of scientific knowledge and offer a glimpse into the deeper, uncharted mechanics of the universe.

The event horizon is a critical threshold in the structure of a black hole, marking the point at which the black hole's gravity becomes insurmountable. Anything—be it light, particles, or entire stars—that crosses this invisible boundary is forever trapped, unable to escape the black hole's gravitational grip. This boundary is what defines a black hole: an inescapable region from which no information can return to the outside world. From the perspective of an outside observer, any object nearing the event horizon appears to slow down and fade, as though frozen in time, due to the intense gravitational time dilation near the black hole.

The event horizon holds a unique significance in our understanding of black holes. Unlike a physical surface, it's a boundary created by gravity itself, marking the transition from the outside universe to a region governed by entirely different physical rules. Once crossed, the laws of relativity tell us there is no way back. Even light, which travels at the highest possible speed in our universe, cannot escape the black hole's gravitational field beyond this point. Thus, the event horizon is often called the "point of no return"—a stark division between the known universe and the unknown.

Within this boundary lies a mysterious and largely incomprehensible region where the black hole's true nature is hidden from view. At the very center of this darkness lies the singularity, a point where matter is thought to be crushed into infinite density. In the singularity, all the mass of the black hole is concentrated into an infinitesimally small point, where gravity reaches unimaginable extremes. The singularity challenges our understanding of physics; here, the fabric of space and time may warp to the point of breaking, with no clear rules governing the behavior of matter and energy. Theories suggest that within the singularity, all known laws of physics collapse, creating a puzzle that has yet to be solved by scientists.

Studying the interior of a black hole, however, is nearly impossible. Because no information can escape from within the event horizon, scientists are left to theorize rather than observe. Our best insights rely on mathematical models and theories that attempt to describe the physics of such an extreme environment. Some theories suggest that black holes could be connected to other regions of the universe or even alternate realities, while others posit that the matter absorbed by black holes simply compresses indefinitely. These ideas, while compelling, remain speculative. The event horizon acts as a veil, hiding the ultimate secrets of black holes and leaving us to wonder about what truly lies beyond.

Chapter 3: Visualization and Imaging of Black Holes

Capturing the first image of a black hole was a groundbreaking achievement that involved an unprecedented level of global collaboration and innovation. This historic image, which revealed the shadow of a black hole in the center of the M87 galaxy, was not captured by a single telescope or location. Instead, it required a global network of radio telescopes working in unison, a project known as the Event Horizon Telescope (EHT). This effort brought together scientists, engineers, and observatories from around the world to create a virtual telescope as large as the Earth itself, a feat that allowed them to achieve the extraordinary resolution necessary to peer across 55 million light-years of space.

The process began with strategically positioning telescopes in key locations around the globe, from Antarctica to the high-altitude Atacama Desert in Chile. Each observatory recorded vast amounts of data, capturing radio waves emitted by the superheated matter swirling around the black hole's event horizon. By synchronizing the timing of each telescope's data collection to within a fraction of a second, the EHT team could effectively combine all the observations, creating an aperture equivalent to the diameter of Earth. This technique, called very long baseline interferometry, allowed them to reach a resolution sharp enough to observe an object billions of times smaller than the galaxy in which it resides.

Processing this data required sophisticated algorithms to combine the various signals and reconstruct a single, coherent image. Even with telescopes working in perfect harmony, gaps in data collection arose due to factors like weather and logistical constraints. To fill in these gaps, researchers employed algorithms designed to interpret and align the scattered data points, essentially "stitching" together a final image that revealed the shadow of the black hole against the glowing disk of superheated gas surrounding it.

This first glimpse of a black hole was more than just a technological feat; it was a visual confirmation of theories that had long predicted what a black hole might look like. The image revealed a dark circle—the shadow of the black hole—encircled by a ring of light, created by the accretion disk of material orbiting at nearly the speed of light. This image brought an abstract concept into the realm of human perception, capturing the public's imagination and validating decades of scientific theory. The success of the Event Horizon Telescope's image underscores both the power of international cooperation and the ingenuity required to push the boundaries of astronomical observation, achieving a view into one of the most mysterious and extreme regions of the universe.

Surrounding a black hole lies a whirling halo of material known as the accretion disk, a structure formed by gas, dust, and stellar remnants that are pulled in by the black hole's immense gravity. As these particles spiral closer to the event horizon, they accelerate to tremendous speeds, heating to millions or even billions of degrees. This extreme heat causes the material in the accretion disk to emit intense radiation, making it one of the brightest regions in the vicinity of a black hole, despite the darkness of the core it encircles. The contrast between the glowing accretion disk and the black hole's shadow allows us to visually identify the location of the event horizon.

One of the remarkable phenomena created by the black hole's gravity is the photon ring, a thin ring of light that encircles the event horizon. The black hole's gravitational pull bends light from the accretion disk, causing some photons to orbit the black hole multiple times before escaping into space. This creates a series of distorted, concentric images of the accretion disk. The photon ring is a result of this repeated light bending, appearing as a luminous circle around the black hole's shadow.

This gravitational lensing effect distorts and magnifies the light, adding layers to the image we perceive and allowing us to glimpse light that originates from behind the black hole.

This effect is more than just a simple ring of light; it's a cosmic mirror that lets us see multiple aspects of the accretion disk simultaneously. Because the black hole's intense gravity warps space-time, light doesn't travel in straight lines around it. Instead, the light curves and bends, creating the illusion that we are viewing the disk from multiple angles at once. Observers looking at a black hole can see not only the side of the accretion disk facing them but also the far side and even the underside, all reflected back due to the gravitational lensing.

This unique cosmic mirage, created by the warping of light and space-time, gives black holes their surreal appearance. The photon ring and accretion disk together reveal the black hole's otherwise invisible presence, and the gravitational lensing produces the eerie effect of seeing what lies around it from all angles at once. This "cosmic mirror" quality challenges our usual perception of space and depth, illustrating just how profoundly black holes distort their surroundings. Gravitational lensing not only enables us to observe these hidden depths but also offers a glimpse into the astonishing effects of gravity at its most extreme, where space itself seems to bend and twist in response to a black hole's pull.

Chapter 4: The Singularities and Spacetime Distortion

At the heart of every black hole lies the singularity, a point where matter is compressed to an unimaginable degree, reaching infinite density. This core is a place where all the mass of a collapsed star converges, crushed by the relentless force of gravity into a space so small that it defies human comprehension. The singularity represents a breakdown in the laws of physics as we know them, where the known equations of general relativity no longer apply. In this extreme environment, concepts like space, time, and matter lose their usual meaning, leading scientists into a realm of theoretical exploration and deep mystery.

What actually happens to matter and energy when they are pulled into the singularity remains one of the most profound puzzles in modern science. As matter approaches the singularity, it's subjected to forces of such intensity that its individual particles, and even the fundamental structure of atoms, are ripped apart. The matter is compacted into an infinitely small point, where density and gravitational force reach levels that cannot be reconciled with the principles of conventional physics. At this scale, gravity and quantum mechanics—the two pillars of modern physics—clash, producing contradictions that scientists have yet to resolve.

Some theories suggest that, within the singularity, all the information about the matter and energy consumed by the black hole is somehow preserved, though in a form we cannot yet understand. Others posit that matter could be compressed into pure energy or reduced to a state beyond anything observed elsewhere in the universe. The singularity challenges our comprehension of what it means to "exist," as all normal dimensions and properties are compressed into this infinitely dense point.

The singularity remains both a source of fascination and frustration for physicists, as it lies beyond the reach of observation or direct study. Current scientific models break down in the face of such extreme conditions, making the singularity a boundary not just in physical space, but in human knowledge. The quest to understand it drives the development of new theories, including quantum gravity and the search for a theory of everything that could reconcile the contradictions between general relativity and quantum mechanics. The singularity at the center of a black hole is more than a point of infinite density—it is a frontier, a mystery that pushes the limits of our understanding of the universe.

Near a black hole, time itself begins to warp in a way that defies our usual experience. This phenomenon, known as time dilation, is a direct result of the black hole's intense gravitational pull, which affects the very fabric of space and time. According to Einstein's theory of relativity, gravity is not just a force but a curvature in space-time, meaning that massive objects like black holes create depressions in the "grid" of the universe. The closer one approaches a black hole, the stronger this curvature becomes, leading to extraordinary effects on time.

For an external observer watching an object approach a black hole, time appears to slow down for the approaching object as it nears the event horizon. If a spaceship, for example, were to move closer to the event horizon, those on the ship would continue to experience time normally. However, to an outside observer, the ship would appear to slow down, seeming almost frozen in place at the edge of the black hole. As the ship gets even closer to the event horizon, it would appear to stop altogether, fading out of view as the light reflecting from it becomes stretched and redshifted, eventually fading into darkness.

For someone moving toward the black hole, however, the experience of time would feel entirely normal, even as the outside universe seems to accelerate around them. This effect grows more intense with proximity to the event horizon. Time passes at a drastically slower rate near a black hole, meaning that a few moments near the event horizon could equate to years, decades, or even centuries passing in the external universe. This idea, while seemingly surreal, is a logical consequence of the way gravity warps space and time.

The intense gravitational field of a black hole doesn't just alter time; it also distorts space, compressing distances near the event horizon and bending the paths of light and matter alike. This compression means that space and time are not separate entities near a black hole; they intertwine, creating a distorted reality where familiar concepts cease to function as expected. Time dilation around a black hole thus becomes not only a fascinating aspect of relativity but also a window into how gravity can alter the fundamental nature of reality. In these regions, the ordinary flow of time loses its meaning, and the boundary between past, present, and future becomes blurred, reshaping our understanding of the universe and hinting at the enigmatic nature of space-time under extreme gravitational conditions.

Chapter 5: Hypothetical Journeys and Alternate Realities

The idea of escaping a black hole invites us into the realm of speculation and thought experiments, pushing the boundaries of known physics. In the established framework of relativity, nothing can exceed the speed of light, meaning that once an object crosses the event horizon of a black hole, it is irretrievably pulled inward. But what if, in an imagined scenario, faster-than-light travel were somehow possible? Such an ability would theoretically allow an object or even a person to defy the gravitational pull of a black hole, possibly enabling an escape from its grasp.

If someone were able to travel faster than light from within the event horizon, the effects of time dilation would introduce fascinating possibilities. From the perspective of this hypothetical traveler, the flow of time inside the black hole is slower relative to the outside universe. Emerging from a black hole after spending even a brief time within its depths could mean arriving in a distant future, as time would have passed far more quickly outside. The journey could span mere moments from the traveler's perspective, while decades, centuries, or more might have elapsed in the external universe. This scenario presents a form of one-way time travel, where the traveler skips forward into the future.

However, beyond this hypothetical escape lies an even stranger idea: the possibility of emerging in a different part of the universe entirely. Some theories propose that black holes might connect to distant points in space or even parallel universes, forming what is known as a wormhole, or Einstein-Rosen bridge. In this model, crossing the event horizon might lead a traveler through a cosmic passage, emerging somewhere far from where they entered. Such an escape would not only mean traveling through time but also traversing immense distances in space, providing an otherworldly "shortcut" that bypasses the normal constraints of travel.

Though the concept of faster-than-light travel remains speculative, it offers a glimpse into what might be possible if the laws of physics as we understand them were somehow circumvented. The prospect of using a black hole as a portal to the future or to an entirely new region of the cosmos fuels our fascination with these enigmatic objects. While current science holds that once inside, escape is impossible, these thought experiments keep the door open to the unknown, sparking wonder about what might one day be revealed in our continuing exploration of space-time's most mysterious realms.

The concept of wormholes, or Einstein-Rosen bridges, takes us deeper into the fascinating possibilities of black holes and spacetime. A wormhole is theorized as a tunnel-like structure that acts as a shortcut through spacetime, connecting two distant points in the universe or even bridging different universes altogether. Proposed by Einstein and physicist Nathan Rosen, the idea is rooted in the same equations of general relativity that describe black holes. In theory, if a black hole's intense gravity could warp spacetime to such an extent, it might also create a passage to another part of the universe—a bridge through spacetime that bends in on itself.

Wormholes, if they exist, could offer unprecedented pathways across the cosmos, allowing travel that bypasses the usual constraints of distance and time. This possibility has led scientists and science fiction enthusiasts alike to wonder if black holes might be entrances to such wormholes, creating a gateway to regions of space light-years away or, perhaps, to entirely different realms of reality. In this scenario, entering a black hole could mean venturing not only into a distant part of our own universe but possibly into a parallel or alternate universe.

This idea feeds into the intriguing mirror universe theory, which speculates that crossing the event horizon of a black hole might lead to a parallel world—a universe existing alongside our own but potentially with alternate laws of physics, altered versions of events, or even completely different landscapes and histories. If such a parallel universe exists, it might resemble our own, mirroring its structure and patterns, or it could be a vastly distinct reality, governed by entirely different rules and constants. These mirror universes, if accessible, could open the door to an endless array of realities, each with its own version of existence.

The potential of encountering a mirror universe through a black hole's event horizon challenges our understanding of reality. If such a journey were possible, it would raise profound questions: Would events play out similarly in this alternate universe, or would we encounter entirely foreign phenomena? Could we find versions of ourselves, or would these universes contain entirely new forms of life and matter? Theoretical though it is, the possibility of mirror universes and alternate realities brings a new dimension to the mystery of black holes, turning them from regions of gravitational collapse into gateways to realms beyond our imagination.

While wormholes and mirror universes remain speculative, these ideas capture the spirit of discovery that drives us to explore the unknown. Black holes might be more than just cosmic endpoints; they could be windows into other dimensions, offering a tantalizing glimpse into the broader and stranger possibilities that might exist within the infinite fabric of the cosmos.

Chapter 6: Black Holes and the Cosmos

At the heart of most galaxies, including our own Milky Way, resides a supermassive black hole, a gravitational giant millions or even billions of times more massive than our Sun. These central black holes are not mere cosmic curiosities; they play a fundamental role in the structure, formation, and stability of galaxies. The supermassive black hole at the center of the M87 galaxy, whose shadow was captured in the first-ever image of a black hole, exemplifies these titanic forces. With a mass approximately 6.5 billion times that of our Sun, it exerts a powerful gravitational influence over the galaxy's core, shaping its structure and evolution.

These supermassive black holes act as anchors within galaxies, their gravitational pull influencing the orbits of countless stars, dust clouds, and other celestial bodies. As surrounding matter spirals toward the black hole, it forms an accretion disk that emits intense radiation and jets of high-energy particles, influencing the energy dynamics of the galaxy. This process, while consuming material, also produces energy that can regulate star formation rates by heating or dispersing gas clouds that might otherwise collapse to form new stars. In this way, supermassive black holes are thought to contribute to a balance, preventing galaxies from growing uncontrollably and promoting an organized, structured environment.

The presence of supermassive black holes in most galaxies suggests a deep, perhaps primordial relationship between these black holes and the galaxies they inhabit. Some scientists theorize that supermassive black holes may have formed early in the universe's history, providing a gravitational nucleus around which galaxies coalesced. Others believe these black holes grew in tandem with their host galaxies, with the two evolving together over billions of years in a feedback loop of growth and regulation.

Beyond the central regions, the influence of a supermassive black hole extends throughout a galaxy. Even at vast distances from the galactic core, stars and star clusters can feel the gravitational pull of the central black hole, aligning their movements with the galaxy's rotational structure. These central black holes not only shape the physical structure of galaxies but also contribute to their stability, ensuring that stars and other celestial bodies remain gravitationally bound, creating the spiral, elliptical, or irregular formations we observe in the cosmos.

In essence, supermassive black holes are far from passive objects; they are dynamic forces that exert control over the galaxies that surround them. Their influence is woven into the very fabric of galactic structure, from their ability to drive matter inward and regulate star formation to their role in holding galaxies together. These gravitational giants at the centers of galaxies remind us that black holes are not only objects of mystery but also essential components in the cosmic architecture, weaving order into the vast tapestry of the universe.

Not all black holes are static entities; many are in fact spinning, a characteristic that adds fascinating complexities to their structure and behavior. These rotating black holes, known as Kerr black holes, differ significantly from their non-rotating counterparts. When a black hole spins, it drags the surrounding space-time along with it, creating a phenomenon known as "frame-dragging." This effect alters the nature of the black hole's event horizon and surrounding regions, producing some remarkable theoretical consequences.

The spin of a Kerr black hole affects its event horizon, slightly flattening it at the poles and widening it at the equator. This rotation generates an additional region outside the event horizon called the ergosphere, where space-time is twisted to such an extent that anything within it, even light, is forced to move in the direction of the spin. Although objects within the ergosphere have not yet crossed the event horizon and can technically escape, they are still subject to the black hole's rotational pull, which bends their paths and distorts their movements.

The spin also affects the behavior of the accretion disk surrounding the black hole. In a Kerr black hole, the intense rotation allows the accretion disk to orbit closer to the event horizon than it could around a non-rotating black hole. This proximity intensifies the heat and speed of particles in the disk, enhancing the emission of energy and radiation. The faster the black hole spins, the closer matter can approach without being immediately pulled into the event horizon, creating an even brighter and more dynamic accretion disk.

Spin introduces another intriguing possibility: it could potentially stabilize a wormhole, allowing for theoretical shortcuts through spacetime. In the highly speculative realm of wormhole theory, the rotation of a Kerr black hole might create conditions where a tunnel, or Einstein-Rosen bridge, could remain open long enough for matter to pass through, potentially allowing travel to distant parts of the universe or even to other universes. The idea remains largely hypothetical, as current physics lacks the framework to confirm or practically apply such a concept. Nonetheless, the implications of a rotating black hole on wormhole stability add another layer of mystery to these already enigmatic objects.

Spinning black holes challenge our understanding of gravity, space-time, and even causality, as their rotation alters the very fabric around them. The complexities introduced by Kerr black holes reveal that black holes are not simply regions of intense gravity but dynamic entities with unique properties shaped by their spin. These spinning giants expand the possibilities for what black holes can be and do, deepening the mystery of their role in the universe and pushing the boundaries of what we know about the nature of space, time, and the potential for connections across vast cosmic distances.

Conclusion

As we look toward the future of black hole research, the mysteries surrounding these cosmic giants promise to drive innovation and discovery. With each new technological advancement and theoretical breakthrough, we edge closer to uncovering the secrets hidden within black holes, yet they remain among the most elusive objects in the universe. Recent progress in high-powered telescopes, such as the Event Horizon Telescope, has already yielded astonishing results, including the first-ever image of a black hole's shadow. Such achievements remind us of how far we've come in understanding black holes and hint at the potential insights that lie on the horizon.

Emerging projects, including next-generation telescopes and advanced detectors, hold the promise of even deeper exploration. Instruments capable of capturing more detailed gravitational wave signals or mapping x-ray emissions from black hole accretion disks may soon allow us to study black holes in unprecedented ways. Upcoming space observatories and innovations in artificial intelligence and data processing are also expected to refine our view, offering sharper images, more precise measurements, and perhaps even a way to observe phenomena within the event horizon. These advancements push the boundaries of observational astronomy, opening windows into the farthest reaches of the cosmos where black holes reside.

Yet, despite our progress, black holes still embody the ultimate mystery, bridging the known with the unknown. They challenge our scientific knowledge, raising questions that venture beyond the realm of physics into the philosophical. What truly lies beyond the event horizon? Could the singularity at the core of a black hole contain answers about the universe's origin, or does it hold realities we may never fully comprehend? The study of black holes forces us to confront the limits of our understanding, creating a space where scientific inquiry meets existential wonder.

Black holes thus serve as a profound reminder of the vastness and complexity of the cosmos. They are not merely astronomical objects but symbols of both our curiosity and the humbling recognition that some aspects of the universe may always lie beyond our reach. As science progresses, we continue to explore these enigmatic entities, hoping that each discovery will bring us closer to understanding the intricate dance of matter, energy, and space-time. In the end, black holes may never fully reveal their secrets, but they will remain an endless source of inspiration and a testament to humanity's quest to understand the deepest mysteries of existence.